LAN Turtle

A Guide to the Covert Remote Access Toolkit

Darren Kitchen

Hak5 Founder

LAN Turtle Creator

HAK5 GEAR

Hak5 LLC, 548 Market Street #39371, San Francisco, CA 94104

LAN TURTLE
A GUIDE TO THE COVERT
REMOTE ACCESS TOOLKIT

International Standard Book Number:
ISBN-13: 978-0-9983732-4-9
Printed in the United States of America
First Printing: November, 2017

Hak5 LLC
548 Market Street #39371
San Francisco, CA 94104

www.Hak5.org

eef5204d6a

CONTENTS

About the Author

Darren Kitchen is the founder of Hak5, the award winning Internet television show inspiring hackers and enthusiasts since 2005. Breaking out of the '90s phone phreak scene, he has continued contributing to the hacker community as a speaker, instructor, author and developer of leading pentesting tools.
darren.kitchen

About the Developer

Sebastian Kinne has led software development at Hak5 since 2011. His background in embedded systems and reverse engineering has been instrumental in the success of the WiFi Pineapple, the popular WiFi auditing tool. As an instructor and speaker on WiFi security, chances are he's sniffed your packets in a demo or two. *sebkinne.com*

About the Editor

Shannon Morse is Hak5's host, producer, and editor. Her focus is on security, technology, and DIY for hackers and consumers alike. Shannon is an advocate for women in tech and she welcomes beginners with her easy to understand tutorials.
snubsie.com

About the Artist

Taz Bedevilled is a South African Illustrator/Graphic Artist. She has a degree in Fine Arts and has worked in the industry since 2012. Aside from freelancing, when she gets a quiet moment she enjoys working on her personal projects, such as the web comic: Avocat & His Guac-Rocket. *tazbedevilled.com*

1
The LAN Turtle

The LAN Turtle is a covert Systems Administration and Penetration Testing tool providing stealth remote access, network intelligence gathering, and man-in-the-middle monitoring capabilities.

Housed within a generic "USB Ethernet Adapter" case, the LAN Turtle's covert appearance allows it to blend into many IT environments.

Use Cases

While the LAN Turtle is exceptionally good at remote access, it also has an open architecture which allows it to excell at other functions. For the most part, it is designed for two primary users

– penetration testers and systems administrators. Some example uses include:

Penetration Testers

- Pivot with a persistent Meterpreter session in Metasploit.
- Scan the network using nmap.
- DNS Spoof clients to phishing sites.
- Exfiltrate data via SSHFS.
- Man-in-the-Middle inline computers capturing browser traffic.

Systems Administrators

- Access to the entire LAN through a site-to-site VPN.
- Automate management scripts with hourly email reports.
- Receive alerts based on defined network traffic
- Maintain out-of-band remote access (3G model)
- Quickly gather pcap (packet captures) files (SD model)

Possible Deployments

The LAN Turtle's discrete nature lends itself to some creative deployments. Here are some for inspiration.

- Covertly installed inside a corporate network to provide the penetration tester with remote access via Metasploit's Meterpreter.
- Surreptitiously installed on a target computer to poison DNS, providing the penetration tester with possible phishing endpoints.
- Secretly installed between a target computer and LAN to intercept and log web traffic.
- Exfiltrate shared data from within the host LAN to an outside SSH filesystem.
- Act as an OpenVPN Gateway in to the host LAN allowing the systems administrator to manage or the penetration tester to audit all nodes accessible to the LAN Turtle remotely.

- Quickly initiate an automated map of the target network and provide a report to the penetration tester either on the device or delivered by email or HTTP.
- Provide a persistent reverse shell over SSH through an intermediate server for remote access to an internal network.
- Masquerade as a legitimate USB Ethernet adapter while performing any automated attack against the host computer or network using tools cross-compiled for the LAN Turtle's Linux-based OS.

Hardware Overview

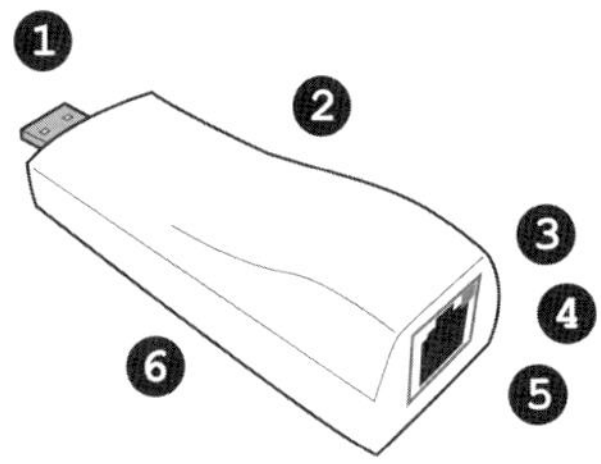

1. USB Ethernet and Power
   Enumerates on the host computer as a generic USB Ethernet device. The onboard Linux-based system uses this Interface (eth0) to offer the host computer an IP address via DHCP.
2. Covert Case
   Generic USB Ethernet Adapter case blends into the target network environment allowing for concealment behind desktops, in telephone closets and server racks.
3. Ethernet
   Standard Ethernet port (eth1) set to obtain an IP address from the host LAN DHCP server. If present, will bridge with a host PC connected to the USB (eth0) port.
4. Green Power Indicator LED
5. Amber Status Indicator LED
6. Reset Button (Inside Case)
   Press to reset to default configuration
   Hold while powering on to access HTTP firmware recovery

Specifications

- Atheros AR9331 SoC at 400 MHz MIPS
- 16 MB Onboard Flash
- 64 MB DDR2 RAM
- 10/100 Ethernet Port
- USB Ethernet Port - Realtek RTL8152
- Indicator LED (Green Power, Amber Status)
- Button (inside case for Factory Reset / Firmware Recovery)
- Dimensions: 95 x 23 x 31 mm

Power Considerations

The LAN Turtle is powered via USB and requires 5V at ~200mA. Typical power usage is 1 Watt. Below are some example deployment scenarios and power considerations.

Covertly installed in an available USB port on the back of a desktop computer at a client site, either in an "Ethernet Pass-through" configuration (With Network access provided by the LAN Turtle) or standalone (such as for DNS Poisoning).

Concealed in a network closet plugged into Ethernet powered by a USB Battery Pack. For example: a large 20000 mAh USB battery would power the LAN Turtle for about 4 days.

Concealed in a telephone room plugged into a free Ethernet cable powered by a typical smartphone USB wall charger.

Concealed in a server rack powered by a server's available USB port using a USB Data Blocker inline USB dongle to prevent the server's operating system from identifying the LAN Turtle.

Connected to a penetration tester's Android tablet or smartphone using a USB OTG cable (available from HakShop.com) for use on-the-go with an Android SSH client. Note: Some smartphones and tablets do not provide the current necessary to power the LAN Turtle. Most Nexus models have been tested to function properly.

2
The Turtle Shell

The LAN Turtle is managed through the Turtle Shell - a text based, menu-driven graphical user interface accessible by SSH. The menus may be navigated using standard arrow, tab, escape and return keys as well as mouse in most terminals.

The Turtle Shell's Configuration Menu provides the ability to change advanced settings such as Password, MAC address, IP address. Firmware updates may be checked for and installed as they become available.

By default the Turtle Shell will start at login via SSH unless disabled from the Configuration Menu.

Exiting the Turtle Shell returns the user to the LAN Turtle's bash shell. To return to the Turtle Shell, run the "`turtle`" command.

Module System

The LAN Turtle uses a modular system for managing its various tools and services. Turtle modules may be started, stopped, enabled, disabled or configured from the Module Menu.

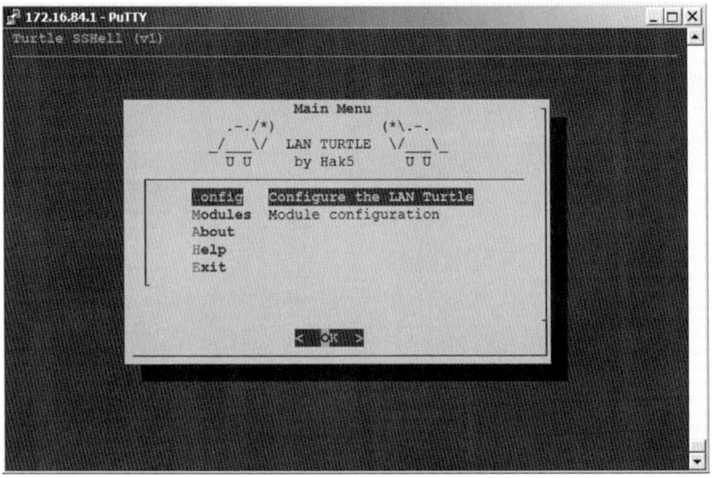

Configure

Configuring a module typically involves entering data specific to your deployment. For example, when configuring the Meterpreter module, you must specify the listening host and port. When configuring the Nmap module you must specify a target, profile and log. Configuration changes made through the graphical Turtle Shell are generally saved in config files and are persistent upon reboot.

Start / Stop

Once configured, a module may be Started or Stopped from the Module Menu. Generally a module may not be started until it has been configured. Some modules, such as the SSH Key Manager, do not support starting and only allow for configuration. Some modules stay running once started. For instance the autossh module will maintain a persistent secure shell until stopped. Other modules will start a task then stop when completed, such as `script2email` or `nmap-scan`.

Enable / Disable

Modules may be set to start up once the LAN Turtle has booted. For example an OpenVPN session may be established upon power-up by first configuring the module and testing it using the start and stop feature. Once the module is achieving the desired result, the module may be Enabled from the Module Menu. Now the LAN Turtle can be deployed on a target network with the OpenVPN module establishing a connection without intervention from the user. An enabled module will start on every boot unless disabled from the Module Menu.

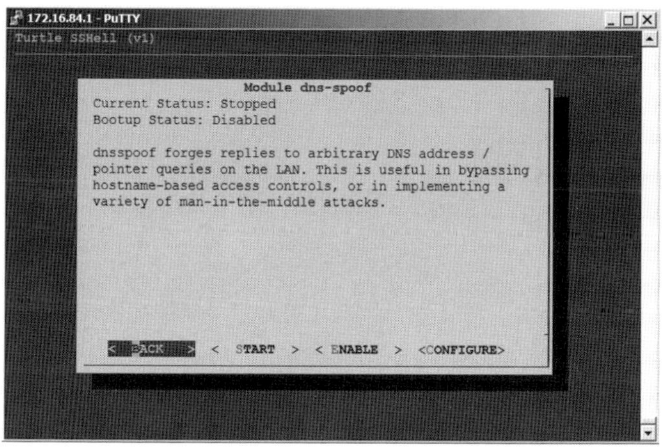

Additional Modules

The LAN Turtle is designed to enable rapid module development. Any supported language (such as bash, php, or python) may be used to write a Turtle Module. With just a few core functions required, modules may be quickly developed and tested. Once submitted to the LAN Turtle Module Repository, they may become available for other LAN Turtle users to download and enjoy. For more information on writing modules, consult the developer documentation at *https://www.LANTurtle.com*

3
Getting Connected

When configuring the LAN Turtle for the first time, a direct connection to the operator's notebook or desktop computer is recommended. The USB plug will both power the LAN Turtle (as indicated by the Green LED) as well as expose a USB Ethernet adapter to the computer for management.

Once connected to the operator's computer via USB, the LAN Turtle will boot. The boot sequence completes in about 30 seconds, during which the the Amber LED will blink. The first time the LAN Turtle is plugged in, the Amber LED will continue blinking until initial configuration is completed via SSH. The very first boot may take a few minutes as the LAN Turtle generates new SSH keys. Subsequent boots should be about 30 seconds.

Once bootup is complete, the LAN Turtle's network interface on the USB facing side will offer the host computer an IP address via

DHCP.

Ensure the host computer is configured to accept IP from DHCP, or alternatively specify a static address in the LAN Turtle's IP range.

Once the LAN Turtle has completely booted and the host computer has been assigned an IP address, the operator may access the LAN Turtle's Shell via SSH.

```
IP Address: 172.16.84.1
      Port: 22
  Username: root
  Password: sh3llz
```

SSH Clients

The LAN Turtle is managed via the Turtle Shell, accessible via SSH. The first time you SSH into the LAN Turtle you'll be prompted to change the root password.

Linux and Mac

Linux and Mac operators are recommended to use the built in SSH client (typically openssh). With the LAN Turtle plugged into the user's PC, an SSH connection is usually initiated by issuing "`ssh root@172.16.84.1`" from the terminal.

Windows

Since Microsoft does not bundle an SSH client with Windows by default, Windows users are encouraged to use the open source PuTTY client from: *http://www.chiark.greenend.org.uk/~sgtatham/putty*

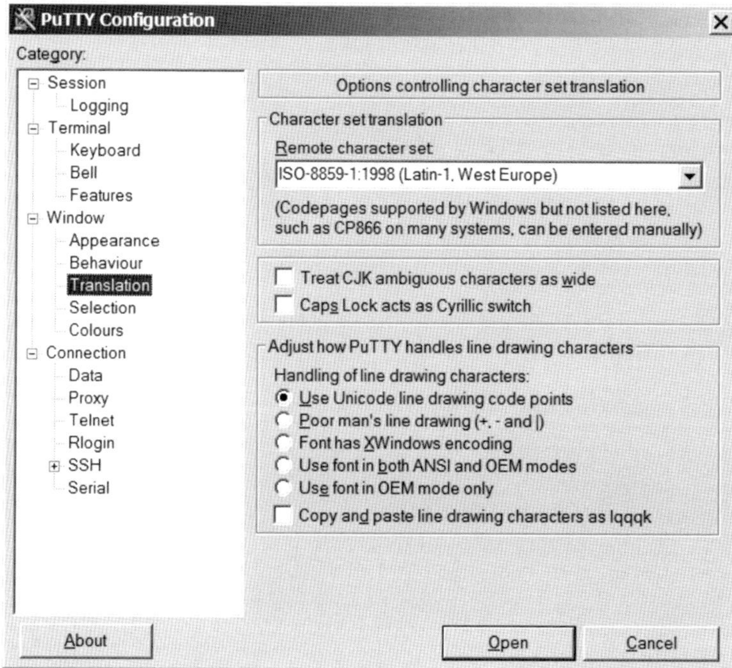

Putty Note:

For the best experience, choose ISO-8859-1 Remote character set from Window > Translation. Otherwise the menu outlines may seem garbled.

Android

For quick access and configuration changes in the field, the LAN Turtle's Shell can be accessed from an Android Tablet or Smartphone by using a USB OTG cable and SSH client. *ConnectBot* is a free open source Secure Shell client available in the Google play store.

4

LAN Turtle Setup

From time to time new version of the LAN Turtle firmware will become available. Thankfully upgrading is made very simple. Begin by getting your LAN Turtle connected to the Internet. Simply connect the RJ45 Ethernet port to a local network. By default the LAN Turtle will attempt to obtain IP information from DHCP. If a static IP address is required on your network it may be set from the Configuration menu.

Test your LAN Turtle's Internet connection. From the SSH session to the LAN Turtle, exit the Turtle Shell menu (or press the ESCAPE key until you're presented with a `root@turtle:~#` prompt). Ping your favorite host.

```
ping -c4 8.8.8.8
```

Once you've verified that your LAN Turtle is online, we're ready to update the firmware. Return to the Turtle Shell by issuing the turtle command. Navigate to the Config screen and choose check for updates. This process takes about 10 minutes to complete. When the LAN Turtle downloads the update and begins flashing, the SSH session will close. This is expected.

When the update is complete you can SSH back into the LAN Turtle. The password will have been reset to the default (`sh3llz`), and once again you'll be prompted to change the password.

Keep in mind that flashing the firmware replaces everything stored on the LAN Turtle memory, so be sure to backup any documents. Modules can be re-downloaded from the Module Manager upon firmware update.

Installing Modules

Installing modules on the LAN Turtle is a simple process. From the Turtle Shell main menu, navigate to Modules. Select Module Manager and press enter. Tab over to select Configure and press enter.

From the module manager configuration screen you'll have the ability to download modules directly from the repository at *lanturtle.com*. Select Directory and press enter, then select Yes to confirm the connection.

The directory will list all of the available modules. Use the spacebar to check the box next to the modules you wish to install, then press enter to continue. Be sure to install the NetCat Reverse Shell (netcat-revshell) module if following along with the next section. Modules will be downloaded from the online repository and installed to `/etc/turtle/modules` on the LAN Turtle.

Similarly, the Delete option from Module Manager allows you to remove any modules you wish. The Update option will get the latest version of all modules currently installed.

Modules may be started, stopped and configured from the Modules menu in the Turtle Shell. Alternatively they may be started, stopped and configured manually from the command line as follows:

```
/etc/turtle/modules meterpreter configure
/etc/turtle/modules meterpreter start
/etc/turtle/modules meterpreter stop
```

5
A Simple Reverse Shell

In this section we'll set up a "backdoor" on the LAN Turtle with Netcat. This will be achieved by configuring a server online to host a Netcat server in the "cloud". On the LAN Turtle we'll configure the Netcat Reverse Shell module to connect to this server every time it starts.

> "Netcat (often abbreviated to nc) is a computer networking service for reading from and writing to network connections using TCP or UDP. Netcat is designed to be a dependable back-end that can be used directly or easily driven by other programs and scripts. At the same time, it is a feature-rich network debugging and investigation tool, since it can produce almost any kind of correlation its user could need and has a number of built-in capabilities. Its list of features includes port scanning, transferring files, and port listening, and it can be used as a backdoor."
> -*Wikipedia*

Begin by SSH'ing into your online server. There are many

inexpensive options for this, and a good VPS or shell is highly recommended. Most VPS hosts come with a static IP address and the operating system of your choice. In this example we'll assume you have an Ubuntu server at 93.184.216.34. From the SSH session with your Ubuntu server online, start a netcat listener on port 8080.

```
nc -lp 8080
```

Netcat is included by default on most Linux distributions. If it is not, try installing it from the repositories. For example, on Ubuntu you may use apt-get to install netcat with the following:

```
apt-get install netcat
```

With our netcast listener running on the server online, we're ready to configure the LAN Turtle. From the SSH session with the LAN Turtle, navigate to the Modules section of the Turtle Shell. Select the Netcat Reverse Shell module and configure. When prompted, enter the IP address and port number of your server's netcat listener. In our example the IP address is 93.184.216.34 and port number is 8080. Tab over to Submit and press enter to save these values.

Now that the Netcat Reverse Shell module is configured, it can be tested by tabbing over to the Start option and pressing enter. You'll receive a notice that the module has started. Press enter to return to the module screen and notice the current status and bootup status.

From the Netcat listener on our server it may not be apparent that anything has happened. This is because the netcat reverse shell does not pass over the prompt. That said, issuing a command will execute on the LAN Turtle just as it would over an SSH connection. Try concatenating the login banner.

```
cat /etc/banner
```

From the SSH session on the LAN Turtle, tab over to the Stop option on the module screen and press enter. You'll receive a notice that the module has stopped. Now go back and notice the netcat listener on the server. It will have terminated. This means we'll be unsuccessful if you attempt to start the module again. To solve this issue, we can either use a version of Netcat with a keepalive option, or run the Netcat listener in a screen session inside of a while loop.

From the server, try the following:

```
screen -dmS netcat_listener bash -c 'while
true; do nc -lp 8080; done'
```

This will create a new detached screen session that will stay persistent even after you disconnect the SSH session with the server. The screen session will be running the bash one-liner `while true; do nc -lp 8080; done`. This simple while loop means as soon as the NetCat listener terminates, it will start again. You can display the running screen sessions with the screen -list command, and reconnect to a running screen session with the `screen -r netcat_listener` command. Detach from the screen session again using the `CTRL+a, d` keyboard combination.

Similar to NetCat, if screen isn't installed by default on your server you may install it from the repositories. For example, on Ubuntu you may issue `apt-get install screen`.

With this more robust listener running, we can now enable the Netcat Reverse Shell module on the LAN Turtle. With the module enabled, the reverse shell will attempt to establish with our server every time the LAN Turtle boots.

Now this is a very basic reverse shell over NetCat. It can be used to manage the LAN Turtle from afar, as long as both you and the

LAN Turtle have access to the server online. It illustrates the basic process of configuring, testing, enabling and deploying a module on the LAN Turtle.

Taking this a step further, you are highly encouraged to set up a persistent reverse shell over SSH since the AutoSSH module is much more robust than netcat, having built-in keepalive features. It can also be used in conjunction with several other features, such as the remote storage module `sshfs`. The process is very similar, requiring configuration of public / private key-pairs from the `keymanager` module.

6

Firmware Recovery and Updates

In the extreme case that a LAN Turtle has become permanently inaccessible or inoperative, there is a quick method for recovery using a special web interface.

Download the latest factory image from *https://downloads.lanturtle.com/* .

Carefully open the LAN Turtle case by removing the two screws underneath the label. While holding the button on the bottom of the LAN Turtle, plug the device into a computer.

After 3 seconds let go of the button and wait an additional 30 seconds to receive an IP address from the LAN Turtle.

If you do not receive an IP address in the `192.168.1.x` range from the LAN Turtle within a minute, statically assign the LAN Turtle's interface to `192.168.1.2` (netmask `255.255.255.0`).

Browse to the LAN Turtle firmware recovery web interface at *http://192.168.1.1.*

Follow the on screen prompts to upload and flash the factory image.

When the flash is complete the LAN Turtle will reboot and will be accessible again from `172.16.84.1` with the default username root and password `sh3llz.`

LAN Turtle Manual Upgrade

LAN Turtle firmware may be updated *over the air* by choosing Check for Updates from the Config menu. If an Internet providing Ethernet connection is not available, updates may be flashed to the device manually using the following process:

Download the latest UPDATE file from *https://downloads.lanturtle.com/.*

Manually SCP the file to the LAN Turtle in /tmp

```
scp turtle-2.bin root@172.16.84.1:/tmp/
```

From the LAN Turtle, exit shell to the bash prompt and issue:

```
sysupgrade -n /tmp/turtle-2.bin
```

Wait about 5 minutes for the LAN Turtle to flash the firmware and reboot.